MW00699791

monday morning®

# Book Factory

When I want to read a good book, I write one.
Benjamin Disraeli

By Murray Suid and Wanda Lincoln

Illustrated by Lisa Levine

This book is for the Hamiltons
and
for Dale Swall

Publisher: Roberta Suid
Editor: Mary McClellan
Design: David Hale
Production: Susan Pinkerton

Other books by the authors: *Editing, Greeting Cards, Letter Writing, Research, Sentences, Speaking and Listening, Stories, Words, Writing Hangups.*

ISBN 0-912107-72-3

Printed in the United States of America
9 8 7

# CONTENTS

# INTRODUCTION

Some people say there's a book inside everyone. We say there's a dozen or a score or maybe too many to count. Our goal is to help you help kids bring out those books.

This doesn't mean turning every kid into an Emily Brontë or a Charles Dickens. No one can teach genius. But we can make it possible for children to experience the joy of publishing their words and pictures.

## WHAT TO DO AND HOW TO DO IT

*The Book Factory* features dozens of projects—everything from ABC primers to yearbooks. You'll find a model page for each activity, a list of easily available materials, and illustrated step-by-step directions. One tip is a must for every successful bookmaking experience: your young authors should always work out their words and pictures first on scratch paper. This is the secret for producing a finished work worth reading.

All of the books can be done individually, and most can also be produced in groups where each child contributes a page: for example, a riddle or a short biographical sketch. The only nongroup books are those that involve continuity from page to page: for example, the Teeny-tiny book.

The Resources section at the back of the book provides a classroom-tested method for producing handsome books on a shoestring budget. In addition to tips on binding, it includes suggestions for making borders, bookmarks, illuminated letters, eye-catching illustrations, and catalog cards.

## FINDING READERS

A book isn't something to be hidden in a drawer. It draws its life from the people who pick it up. Therefore, the culminating activity should be to circulate each book. This can mean anything from giving it as a gift to putting a copy on the library shelf (with a card in the card catalog) to displaying the book in the window of a local bookstore. (See Resources.)

Of course, the most important reader for these books will be the author. We hope that the children will experience the kind of pleasure that Sir James Barrie felt upon the publication of his first book:

> I carried it about in my pocket and took surreptitious peeps at it to make sure the ink had not faded.

# ABC THINGS

**MATERIALS:** paper, writing tools, dictionary, tiny objects, magazines or newspapers, scissors, glue or stapler or tape, plastic bag

**DIRECTIONS:**
1. For each letter of the alphabet, find a small object. The object should be thin enough to fit between the pages of a book: for example, a comb.
2. If an object can't be found for some letters, draw a picture or cut one from a magazine or newspaper.
3. Attach each object or picture to a page using glue or a stapler or tape. Or put the object into a plastic bag and attach the bag to the page.
4. Label the object and include other words that fit the letter.

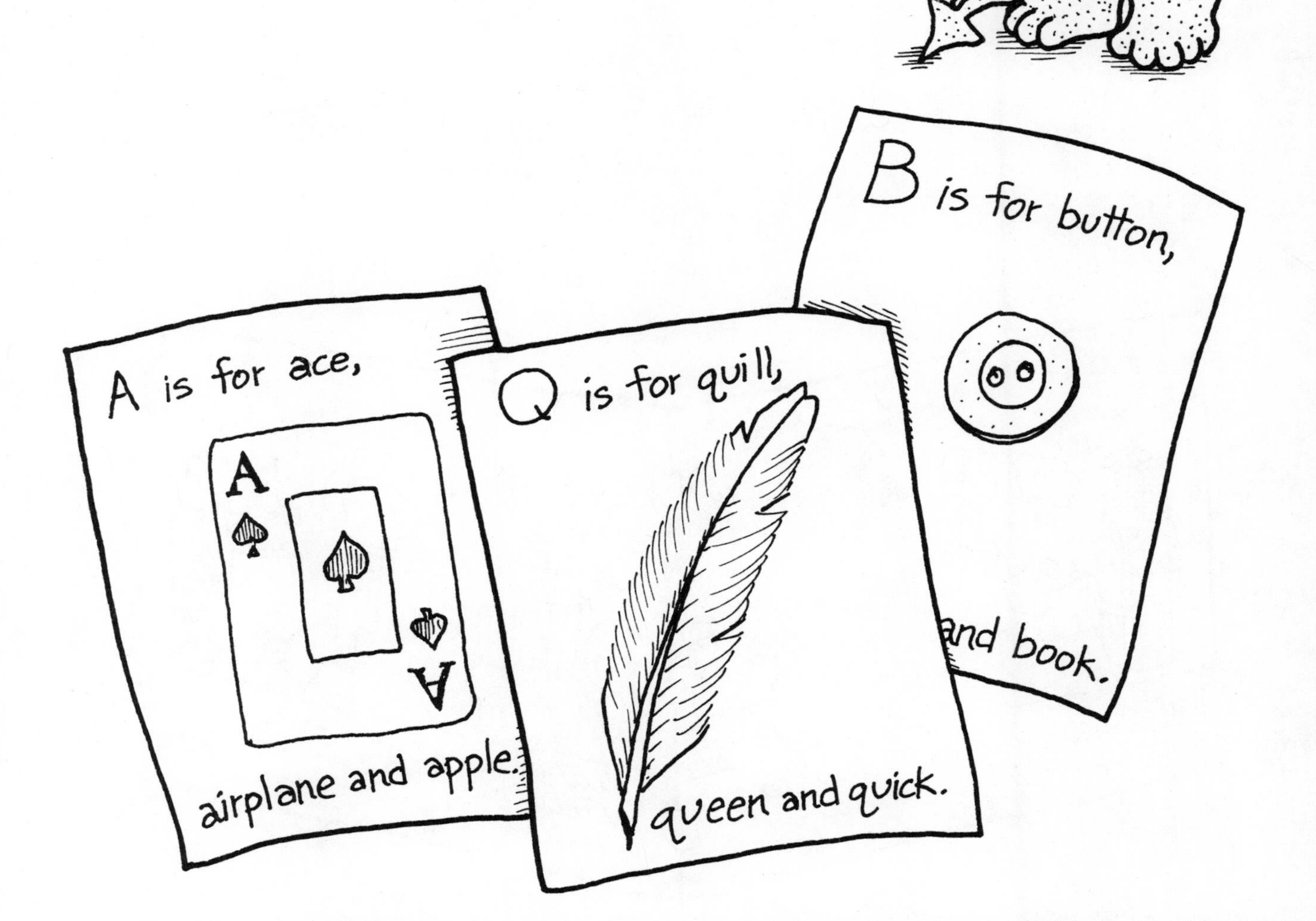

A is for ace,
airplane and apple.
Q is for quill,
queen and quick.
B is for button,
and book.

# ACCORDION BOOK

**MATERIALS:** paper, writing tools, tape

**DIRECTIONS:**
1. Plan a book that tells the story of a field trip or other adventure.
2. On a separate page, draw a picture for each event in the trip.
3. Write the words that go with each picture.
4. Put the pages in order. Then lay them out side by side on a long table or on the floor.
5. Use transparent tape to join each picture to the one on its left and right.
6. Next, turn over the entire strip and tape the other sides.
7. Finally, fold the book like an accordion.

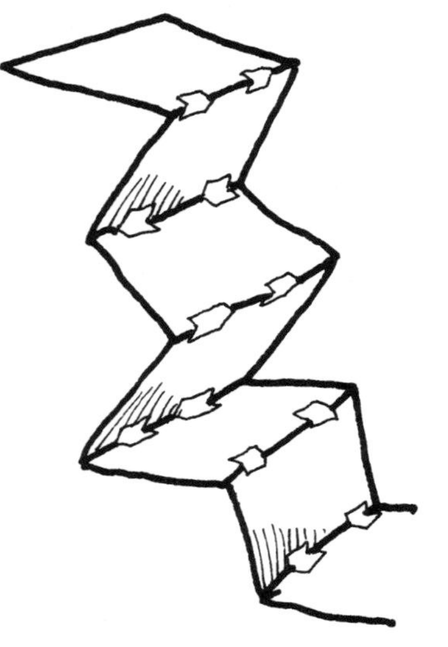

# ANIMAL READER

**MATERIALS:** paper, writing tools

**DIRECTIONS:**
1. Imagine a world in which birds, kittens, worms, dogs, and other animals could read. What kinds of books would they like?
2. Pick an animal, and think up a book idea that would interest the animal. It could be a storybook, a fact book, a book of poems, or an ABC book.
3. Write the words for each page of the book.
4. Add pictures and maybe a border.
5. Give the book an interesting title.
6. Read the book to the animal or share the book with human readers.

# BABY BIOGRAPHY

**MATERIALS:** paper, writing tools, a photograph of a baby, a current photograph of the same person, scissors, tape or glue

**DIRECTIONS:**
1. Attach the photograph of the baby near the top of the page.
2. Under the picture, write a few sentences that tell about the person when he or she was a baby. This might include favorite foods or favorite activities.
3. Down the page, attach the current photograph.
4. Cover the current photo with a flap.
5. Invite the reader to guess who the person is.
6. Create a book of these photo-riddles—one for everybody in a class or in a family.

*Variation*: Make a baby biography of a famous person.

Who is this person?
This baby used to wake up every night. To get her back to sleep, her dad had to rock her. Hint: Her name rhymes with hill.
Lift to see the person today

# BLANKETY-BLANKS

**MATERIALS:** paper, writing tools, dictionary

**DIRECTIONS:**
1. Write a true or a made-up story. Leave space between each line of the story.
2. Underline one or two words in each sentence.
3. Under each underlined word, write the word's part of speech: for example, noun or adjective. For help, use a dictionary.
4. Copy the story onto fresh paper, but leave out the underlined words. Draw a line where each underlined word was. Under the line, print the missing word's part of speech.
5. Find someone who knows the different parts of speech. Without letting the person see the page, ask him or her to give a word for each blank. Write the words in the blanks.
6. Now read the story aloud. It should be very funny.

*Variation*: Instead of a story, use a poem.

# CHOOSE-THE-ENDING STORY

**MATERIALS:** paper, writing tools, Choose-the-Ending Story Planner

**DIRECTIONS:**

1. Think up a story that puts the main character or characters into a lot of tricky situations: for example, being chased by a monster.
2. Write the first page of the story. Add a picture for interest. The page should end with the character having two clear choices. Explain the choices by writing something like this:

   If you think Isaloo should jump on the bicycle, turn to page 2. But if you think Isaloo should climb the tree, turn to page 3.
3. Next, write page 2. At the bottom of that page, give two more choices. Do the same for page 3.
4. Continue the story for about three branches, and complete the book by writing the different endings.
5. Bind the book and give it an exciting title and cover.

THE ADVENTURES OF LILLABELL.
by
CHOOSE-THE-ENDING!


*Hint*: To keep track of the different choices, use a story planner like the one shown on the following page.

Suddenly, out of a big black fin!
If you think Lillabell should try to swim away from the shark, turn to page 16.
If you think she should to make friends the shark,
17.

## Choose-the-Ending Story Planner

In a Choose-the-Ending Story, each page gives the reader two choices at the bottom of each page. The choices are about which page the reader should turn to. The choices continue until the reader is sent to a page that ends the story.

The following list will help the writer figure out which choices to give the reader at the bottom of each page.

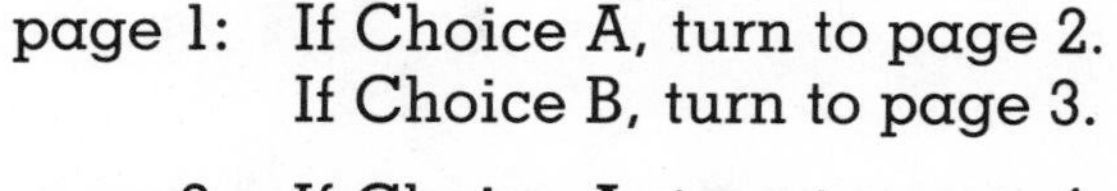

page 1: If Choice A, turn to page 2.
If Choice B, turn to page 3.

page 2: If Choice A, turn to page 4.
If Choice B, turn to page 5.

page 3: If Choice A, turn to page 6.
If Choice B, turn to page 7.

page 4: If Choice A, turn to page 8.
If Choice B, turn to page 9.

page 5: If Choice A, turn to page 10.
If Choice B, turn to page 11.

page 6: If Choice A, turn to page 12.
If Choice B, turn to page 13.

page 7: If Choice A, turn to page 14.
If Choice B, turn to page 15.

page 8: If Choice A, turn to page 16.
If Choice B, turn to page 17.

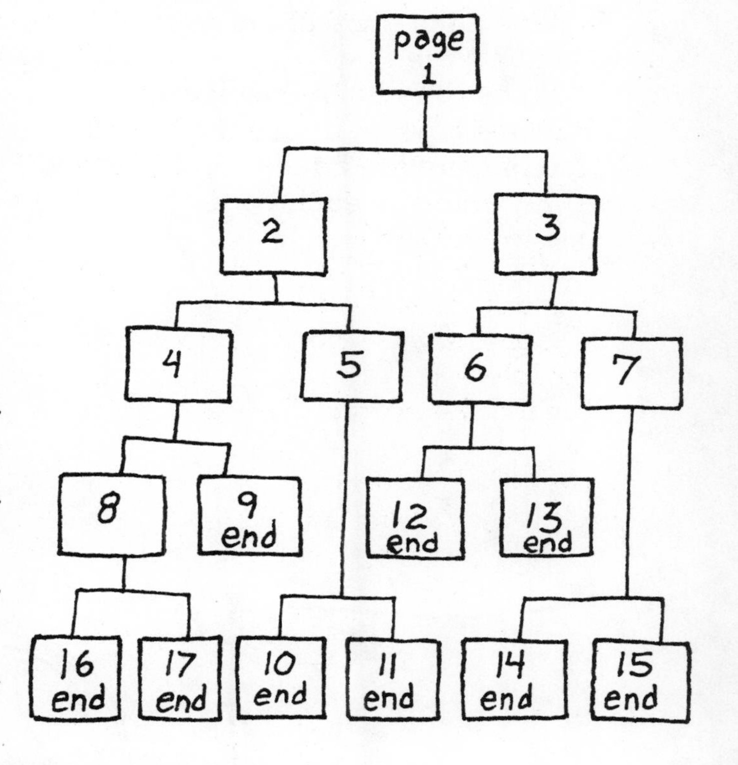


At this point, write nine separate endings, one for each page from 9 through 17. In other words, page 9 will present one ending, page 10 will present another, and so on.

For a longer book, continue the pattern. That is, page 9 will send the reader to either page 18 or page 19. But beware: The more choices, the more endings.

# CLOCK STORYBOOK

**MATERIALS:** paper, writing tools, brads, scissors

**DIRECTIONS:**
1. Think up a story that takes place in a room with a big clock. The passing of time should be important in the story. For example, the characters could be waiting for someone to come home at a certain hour.
2. Write each page of the story. Draw a picture of the clock on each page.
3. Cut out hands for the clocks.
4. On each page, attach the hands to the clock using a brad, so that the hands can be moved.
5. The words on each page should tell the reader what time to set the clock to.

The radio said
By six it would snow.
But it was now 6:30
And not a flake did show
(Move the hands of the clock to 6:30)

# COLORING BOOK

**DIRECTIONS:**

1. Choose a subject for a coloring book page. It could be a person, place, thing, or activity.
2. Draw an outline of the subject.
3. Add a few words or a poem.
4. Put together a book of coloring pages, and then give the book to a young child for some reading and coloring fun.

We need to water to keep the plants growing.

# DICTIONARY

**MATERIALS:** paper, writing tools

**DIRECTIONS:**
1. Pick a familiar subject, such as cooking, music, or a sport.
2. List the important words people use when talking about that subject.
3. Write a definition for each word on the list.
4. In alphabetical order, copy the words and definitions onto fresh paper. When necessary add pictures for a clearer definition.

*Variation*: Make a Dictionography—a dictionary of words about a particular person. These words might include the person's hobbies, job, and favorite possessions.

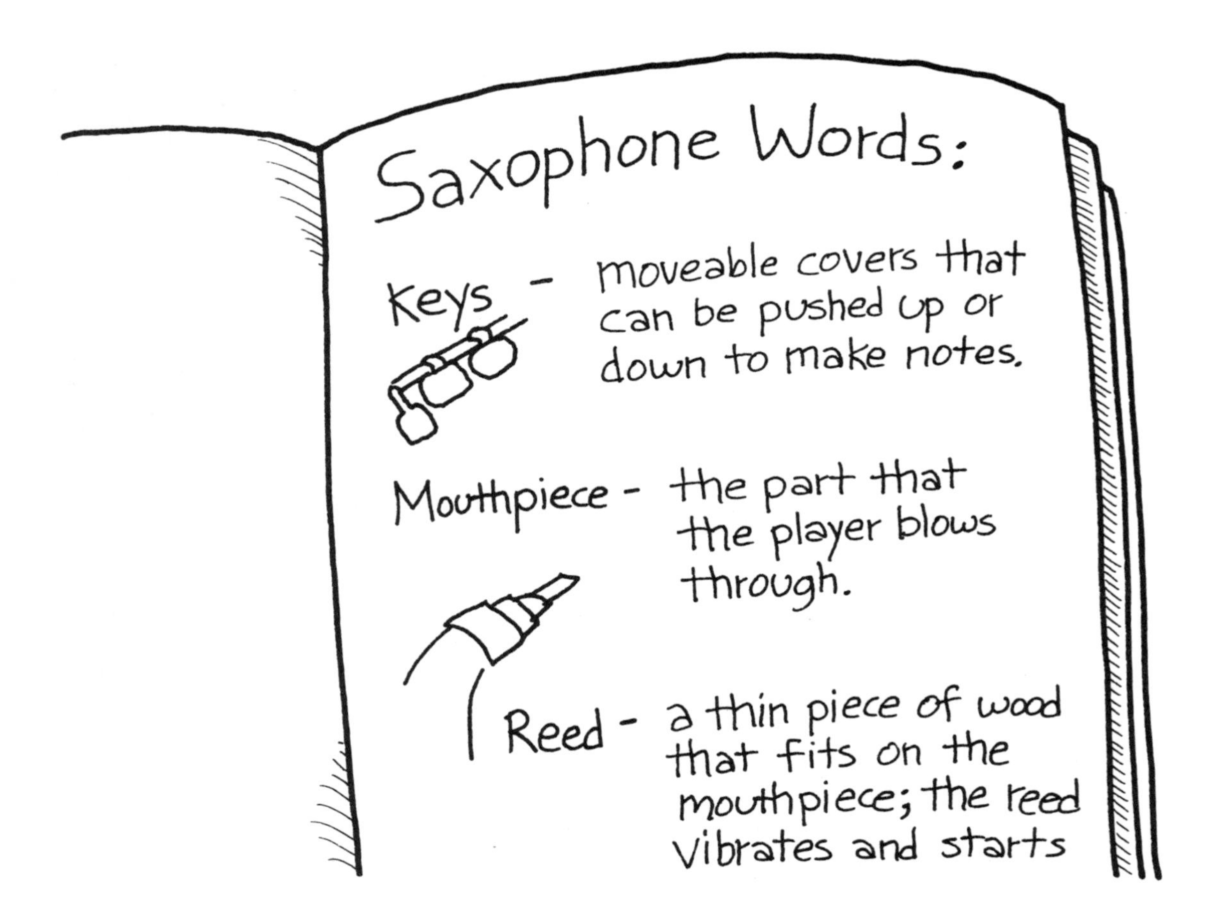

Saxophone Words:
Keys - moveable covers that can be pushed up or down to make notes.
Mouthpiece - the part that the player blows through.
Reed - a thin piece of wood that fits on the mouthpiece; the reed vibrates and starts

# FACT BOOK

**MATERIALS:** paper, writing tools

**DIRECTIONS:**
1. Pick an interesting subject, such as shoes or insects.
2. Study the subject and take notes. Use measuring equipment if necessary. Add facts by reading books or talking to an expert.
3. Write the facts. Add pictures for interest.
4. Collect other fact pages, and make a big trivia book.

Facts About
Bicycle Spokes
by Candace Delone
On most ten-speed bikes, there are 36 spokes.
The average spoke is 30cm (11 3/4") long.

# FLIP BOOK

**MATERIALS:** small notebook or stapled-together index cards, writing tools

**DIRECTIONS:**
1. Get a small notebook, or staple together about a dozen index cards.
2. On the last page of the booklet, make a simple drawing of something that can move: for example, a ball or a flag.
3. On the next page, repeat the drawing, but move the subject a little bit.
4. Keep doing this on each page. From time to time, flip the pages to check the movement.
5. Use the top page for the title.

# HANDY BOOK

**MATERIALS:** paper, writing tools

**DIRECTIONS:**
1. Trace around someone's hand.
2. In the palm space, write the person's name.
3. In each finger, write something special that the person does with the hand: for example, "Plays the piano." Try to list something that most people don't do.

*Variation 1*: Pick someone famous: for example, a hero from history, a sports star, a fictional character, or a TV performer. Draw the person's hand, and fill in his or her handy achievements.

*Variation 2*: Make a foot book. On each page, draw around someone's foot. Inside the foot, tell how the foot is used—for example, to kick a football—or where the foot has been.

# HIDDEN RHYMES

**MATERIALS:** paper, writing tools, scissors, paste or tape

**DIRECTIONS:**

1. Write a rhyme.
2. Draw a picture to show what's going on in the rhyme.
3. Cut a piece of paper that is big enough to cover the last line of the rhyme. Paste or tape the flap over the last line.
4. Invite the reader to guess the line.
5. Tell the reader to lift the flap in order to see the last line of the rhyme.

# HOLIDAY BOOK

**MATERIALS:** paper, writing tools, list of holidays and festivals

**DIRECTIONS:**
1. Pick a holiday to celebrate.
2. Write a story or a poem about what the holiday means or why it is important.
3. Add art.
4. Collect holiday pages from other writers, and bind the pages into a book.

# HOMONYM RIDDLES

**MATERIALS:** paper, writing tools, list of homonyms

**DIRECTIONS:**

1. Pick a pair of homonyms, for example, sale/sail, and tail/tale.
2. Draw a picture that is about these sound-alike words.
3. Under the picture ask the reader to guess the phrase.
4. Print the answer upside down at the bottom of the page or on the back of the sheet.

## Homonyms List

aisle/isle
ant/aunt
ate/eight
ball/bawl
beach/beech
bell/belle
blew/blue
boarder/border
boy/buoy
cents/scents
Chile/chili
colonel/kernel
dear/deer
fair/fare
fir/fur
flea/flee
flew/flue
flour/flower
gnu/new
grease/Greece
hair/hare
hall/haul
heard/herd
hi/high
him/hymn
hoarse/horse
hoes/hose
knight/night
main/Maine
miner/minor
moose/mousse
muscle/mussel
oar/ore
pair/pear
rain/rein
ring/wring
rose/rows
sail/sale
scull/skull
sea/see
son/sun
stair/stare
stake/steak
steal/steel
tail/tale
tern/turn
tide/tied
toad/towed
waist/waste
wait/weight
warn/worn

# HOW-TO-DO-IT BOOK

**MATERIALS:** paper, writing tools

**DIRECTIONS:**
1. Pick an activity: for example, doing a magic trick, using a camera, or solving a math problem.
2. On scratch paper, describe each step in doing the activity. Add pictures to make the directions easier to follow.
3. Try the directions on a reader. If the reader gets confused, change the directions to make them clearer.
4. Copy the finished words and pictures onto one or more pages of a book of directions.

HOW TO DRAW A FACE
1. Make an egg shape.
2. Add ears halfway down the sides of the head.
3. Add a nose.
4. Under the nose, draw a mouth.
5. Above the nose, draw the eyes.
6. To change the face, move the pupils.

## More Ideas for How-To-Do-It Books

How to...

Become a magician
Cheer up a friend
Dive without belly flopping
Do a science experiment
Draw comics
Eat spaghetti
Eat with chopsticks
Fly a kite
Handle a bully
Housebreak a puppy
Improve the school
Improve your memory
Juggle
Kick a football
Knit
Make a friend
Make a paper airplane
Memorize spelling words
Never be bored
Overcome stage fright
Play chess
Play the harmonica
Program a computer
Read a map
Shoot free throws
Study
Take good photographs
Teach someone to ride a two-wheeler

# LIST BOOK

**MATERIALS:** paper, writing tools

**DIRECTIONS:**

1. Think up a title for a list: for example, "Scary Dreams" or "Terrible Foods."
2. List as many items as possible for the list.
3. Illustrate the most interesting item on the list.
4. Give the list a title.
5. Combine the list with lists on other subjects to make a book of lists.

Three unusual places to have a picnic:
1. In a rowboat in the middle of a pond.
2. In the back of a parked pickup truck.
3. Under the dining-room table.
COOLER

# MATH BOOK FOR LITTLE KIDS

**MATERIALS:** paper, writing tools, scissors, tape or glue

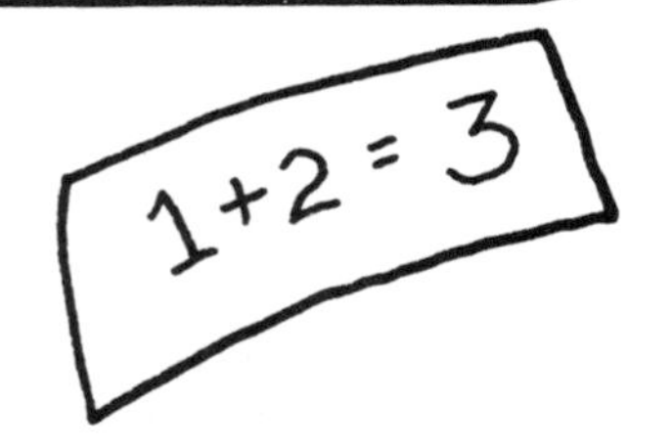


**DIRECTIONS:**
1. Think up a math problem: for example, 2 + 3 = 5.
2. Make a page or two that a little kid could use to solve the problem. For example, a picture could show two ducks in a pond who are joined by three more ducks.
3. Ask the reader to solve the problem.
4. Give the answer on the page, but hide it under a flap.

# MOVEABLE ART BOOK

**MATERIALS:** paper, writing tools, felt, scissors, glue, plastic bag

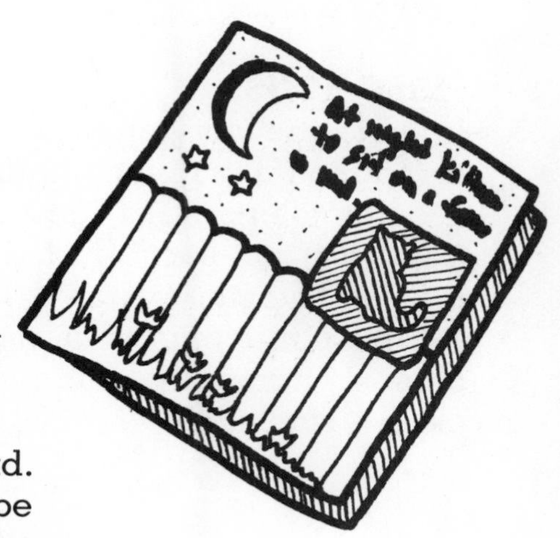

**DIRECTIONS:**

1. Think up a story with something that will appear on most pages, for example, a bird or a car.
2. Cut that thing out of felt.
3. Print the words and draw a picture for each page.
4. Place a piece of felt wherever you want the felt thing to go.
5. Attach a plastic bag to the front of the book. The felt thing will be stored there when the book isn't being read.
6. Write a note to the reader saying that the thing is to be put in the right place as each page is read. It will then be removed when the page is turned.

# NAME BOOK

**MATERIALS:** paper, writing tools

**DIRECTIONS:**
1. Learn the story of someone's name. The name could belong to the writer, a family member, or a friend. It might even be the name of someone famous. Get the facts by talking to someone or reading a book.
2. Write the story.
3. If possible, include a picture that helps explain the person's name: for example, a picture of whomever the person was named for.

# NOISY BOOK

**MATERIALS:** paper, writing tools

**DIRECTIONS:**
1. Think up a story that has lots of sounds in it: for example, whistling kettles, howling winds, ringing church bells, roaring cars, ticking clocks, clapping people, or noisy animals.
2. Write each page of the story. Wherever a sound is needed, tell the audience what sound to make.
3. Before reading the book aloud, have the audience practice each sound.

Now let me hear you honk.
Honk! Honk!
Honk!
Dino heard many strange sounds -- a siren...
(Go "Wheeeeeeeeee!")
cars honking...
(Go "Honk, honk, toot, toot")
But the strangest sound came from people who saw Dino...
(Go "Yiiiiiiiiiikes!")

# OBSERVATION BOOK

**MATERIALS:** paper, writing tools, scissors, tape or glue, magazine or newspaper (optional)

**DIRECTIONS:**

1. Find or draw a picture with lots of things to look at. For example, it might be two leaves with different shapes or a photograph of a city scene clipped from a magazine.
2. Tape the picture onto a page.
3. Ask the reader to study the picture closely and to learn as much about it as possible.
4. Somewhere on the page, tell all of the things that are interesting about the picture. Then cover the words with a flap.
5. Invite readers to lift the flap when they are done looking at the picture. This way they can compare their observation skills with those of the writer.

How are these two leaves alike?
How are they different?
Lift this flap to learn more about these two leaves.

# OPEN IT BOOK

**MATERIALS:** paper, writing tools, scissors, tape

**DIRECTIONS:**
1. Think up a story in which the characters are always opening up doors, curtains, boxes, and so on.
2. On each page, draw what the reader will see when something is opened: for example, the inside of a closet.
3. Cut out the part that covers the opening: for example, a closet door.
4. Tape the covering part onto the page so the reader can open it.
5. Write words that invite the reader to open the thing.

*Variation*: Each page can stand by itself. The words might be a poem or a few paragraphs which make the reader want to see inside the thing.

One day I told my mom
And I also told my dad
"Don't look inside my
Closet, please, 'cause
Something really bad
Is lurking right in there.
An awful monster maybe
Or at least a scary,
Scary, scary..."

# PATTERN WRITING

**MATERIALS:** paper, writing tools

**DIRECTIONS:**
1. Pick a story, poem, or a song that has a well-known pattern: for example, "The Twelve Days of Christmas."
2. Write a new story, poem, or song that uses the old pattern but that has new words.
3. Add pictures to show what's happening in the story or song.

The Many Days of School
BOOKS
On the first day of classes,
My teacher gave to me
A lesson about geography
On the second day of c
My teacher gave to m
Two math worksheets
And a lesson about ge

# PULL-UP RIDDLE BOOK

**MATERIALS:** paper, writing tools, scissors, stapler

**DIRECTIONS:**
1. Find or write a riddle.
2. Draw a picture that illustrates the riddle. Write the riddle at the top of the page.
3. Cut a strip of paper. At one end write: "Pull." Then print the answer to the riddle on the strip.
4. Staple a tab to the answer strip.
5. Cut a slit in the drawing, and slip the answer strip partway through it.

# QUOTATIONS BOOK

**MATERIALS:** paper, writing tools

**DIRECTIONS:**
1. Pick a topic: for example, cats, flying saucers, friendship, homework, money, music, reading, TV, or school.
2. Ask as many people as you can to give their ideas about the topic. Carefully write down what each person says.
3. Print the quotations you like best on a page for a quotation book. Be sure to include the person's name next to his or her quotation.
4. Draw or find a picture that has something to do with the topic.

*Variation*: Make up a pretend-quotations book in which animals, flowers, or things give their opinions.

# REAL PERSON BOOK

**MATERIALS:** paper, writing tools, photographs (optional), scissors, paste

**DIRECTIONS:**
1. Make up a story in which a friend or a relative plays a part.
2. Use facts about the person throughout the story.
3. If possible, paste photos of the person onto the pages. The photos can be combined with artwork.

*Variation*: Put the person into a fairy tale or other old story.

# SCROLL STORY

**MATERIALS:** paper, writing tools, ribbon or yarn

**DIRECTIONS:**
1. On scratch paper, plan a story that has lots of scenes.
2. Copy the story onto a long strip of paper.
3. Add pictures.
4. Roll up the story and tie a ribbon around it.
5. On the outside of the scroll, write the title and the author's name.

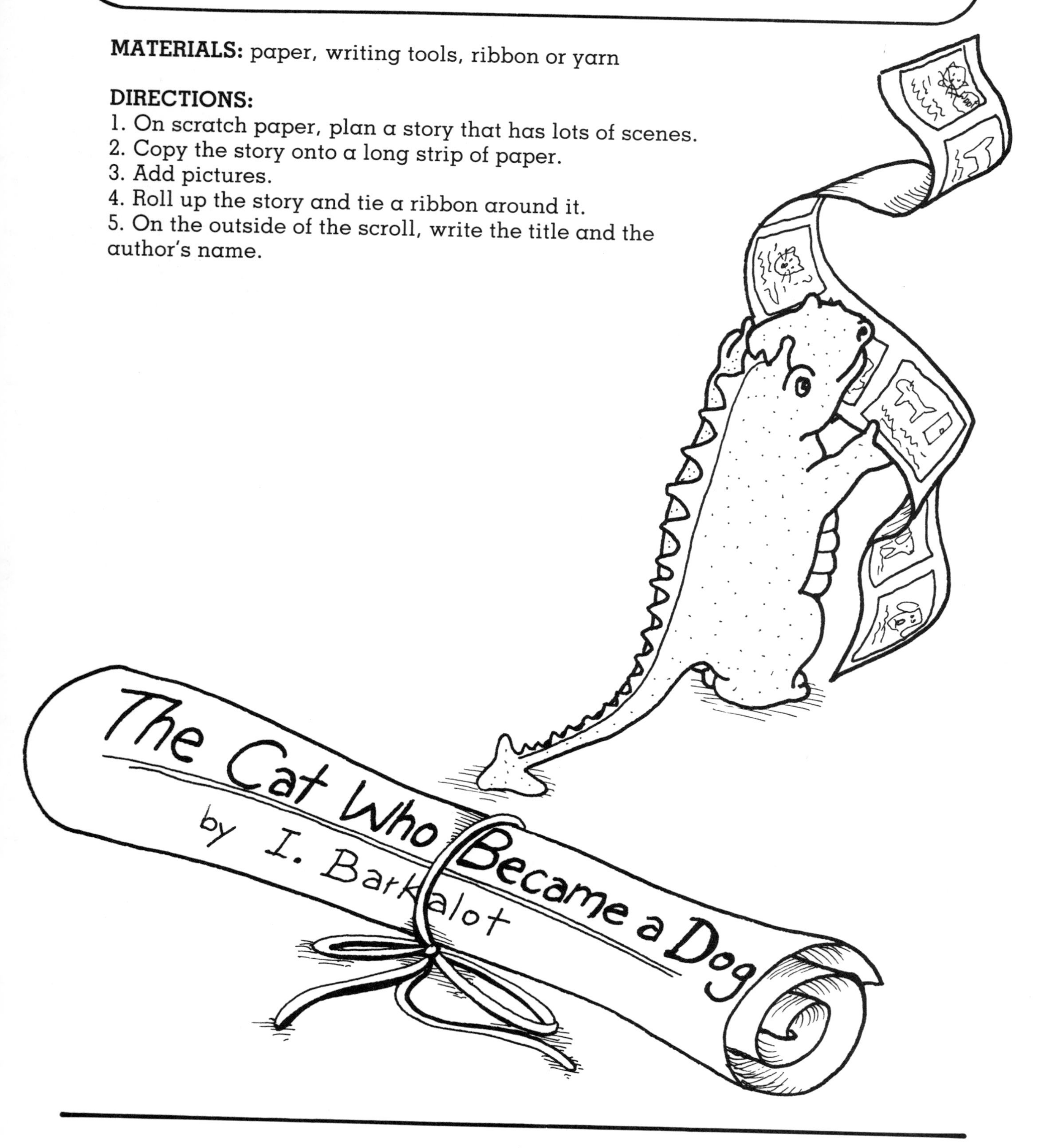

The Cat Who Became a Dog
by I. Barkalot

# SEE-THROUGH GUESSING GAME

**MATERIALS:** paper, writing tools, newspaper or magazine, glue or paste, scissors

**DIRECTIONS:**

1. Cut a picture of a familiar person, place, or thing from a newspaper or magazine, or draw a picture.
2. Paste the picture onto a piece of paper.
3. Cut a small hole in another piece of paper so that when the two sheets are laid together only a small part of the picture will show through the hole.
4. On the piece of paper that contains the hole, write a clue about what the picture shows.
5. Cover the clue with a paper flap that is glued to the sheet.

# SEEDLES

**MATERIALS:** paper, writing tools, seed, tape or glue, plastic bag

**DIRECTIONS:**
1. Get a seed. It could be from an orange, a cucumber, a pumpkin, or a flower.
2. Write a paragraph about the kind of plant that grows from this kind of seed, but don't mention the name of the plant.
3. Tape or glue the seed to the page. If it's a big seed, place it in a plastic bag and attach the bag to the page.
4. Ask the reader to guess what the seed is.
5. On the back of the page, draw a picture of the plant's fruit or flower.

# SHAPE BOOK

**MATERIALS:** paper, writing tools, scissors

**DIRECTIONS:**
1. Draw an object with a well-known shape. Examples include a baseball, key, shoe, telephone, tree, or truck.
2. Cut out the shape and use the pattern to make pages for the book.
3. On each page, write a paragraph or a poem about the object. The words might tell how it is made or used.
4. Make a title page that tells what the book is about.

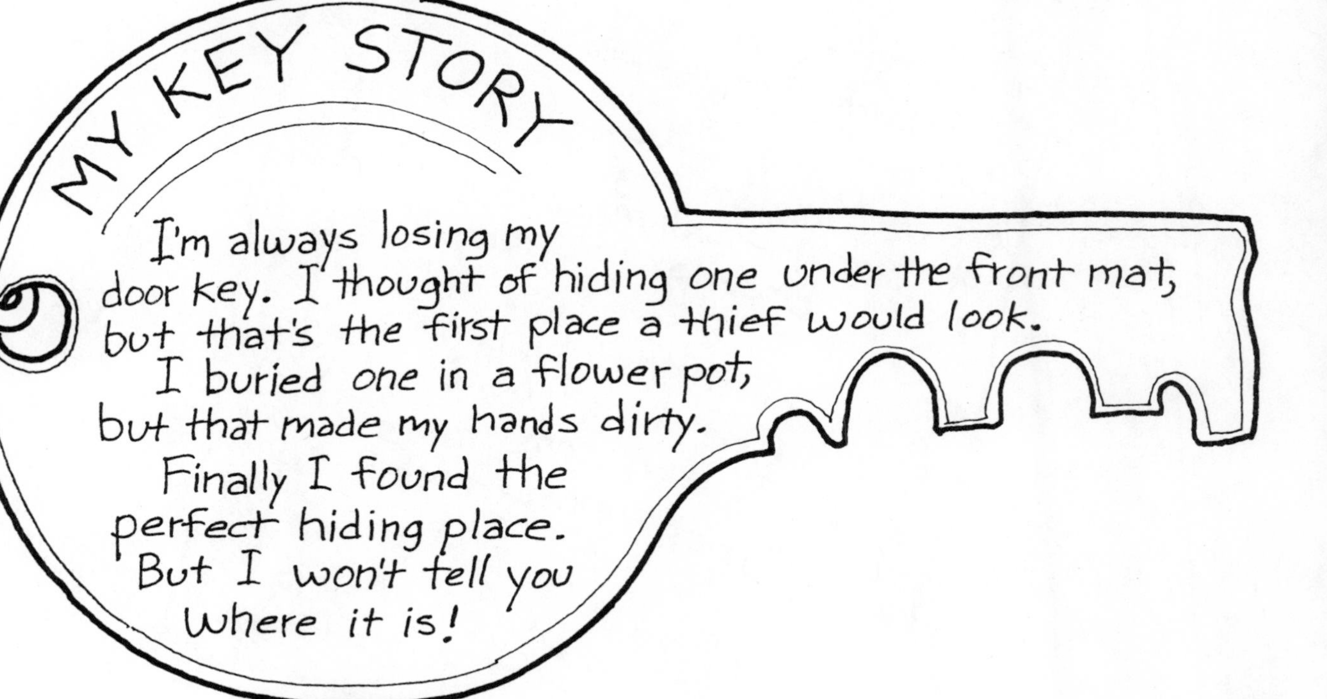

MY KEY STORY
I'm always losing my door key. I thought of hiding one under the front mat, but that's the first place a thief would look.
I buried one in a flower pot, but that made my hands dirty.
Finally I found the perfect hiding place.
But I won't tell you where it is!

# SING-ALONG BOOK

**MATERIALS:** paper, writing tools

**DIRECTIONS:**
1. Pick a subject for a song: for example, riding a bike or watching TV.
2. Choose a melody that most people have heard of. Examples include "Happy Birthday to You," "Mary Had a Little Lamb," "Pop Goes the Weasel," "Row, Row, Row, Your Boat," and "Twinkle, Twinkle, Little Star."
3. Work out a rhyme that fits the melody. Give it a title and tell the reader what tune to use when singing the song.
4. Put a musical border around the page.
5. Make a collection of new songs for old melodies.

STUDY
To the tune of "Twinkle, Twinkle, Little Star"
Study, Study,
All day long.
That will make your
Brain grow strong.
With that brain
One thing to do
Is learn some facts
That are new to you.

# STORY STARTERS

**MATERIALS:** paper, writing tools, envelope, paste or tape

**DIRECTIONS:**
1. Write an exciting first page for a story.
2. Add a picture.
3. Paste or tape an envelope onto the page.
4. On the envelope, print a message that invites the reader to complete the story. Each reader will leave his or her story in the envelope for other readers to enjoy.

*Variation*: Instead of using an envelope, attach blank pages after the story. Ask each reader to complete the story using one of the blank pages.

# TEENY-TINY BOOK

**MATERIALS:** paper, writing tools, scissors, stapler

**DIRECTIONS:**
1. Think up an idea for a book that will have 16 tiny pages. Each page will have a picture and a few words.
2. Fold a piece of paper in half; then fold it in half again but this time across the crease. Do this two more times.
3. Unfold the paper, and cut along the creases. The result should be 16 tiny pages.
4. Fill up each page with a picture and words.
5. Staple the pages.

*Variation*: Prepare the book on regular sheets of paper and then reduce the book in size using a copy machine.

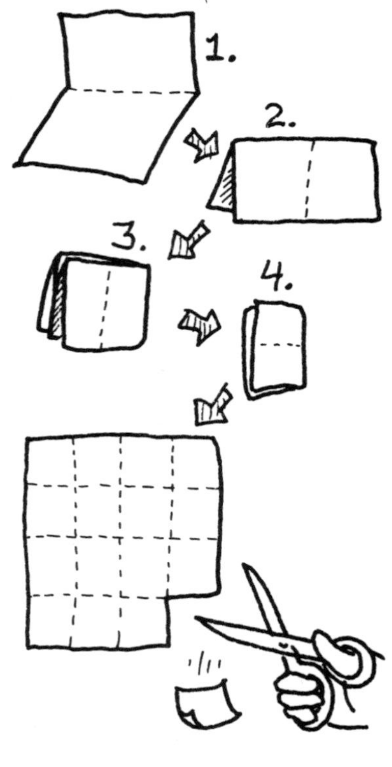
1.
2.
3.
4.

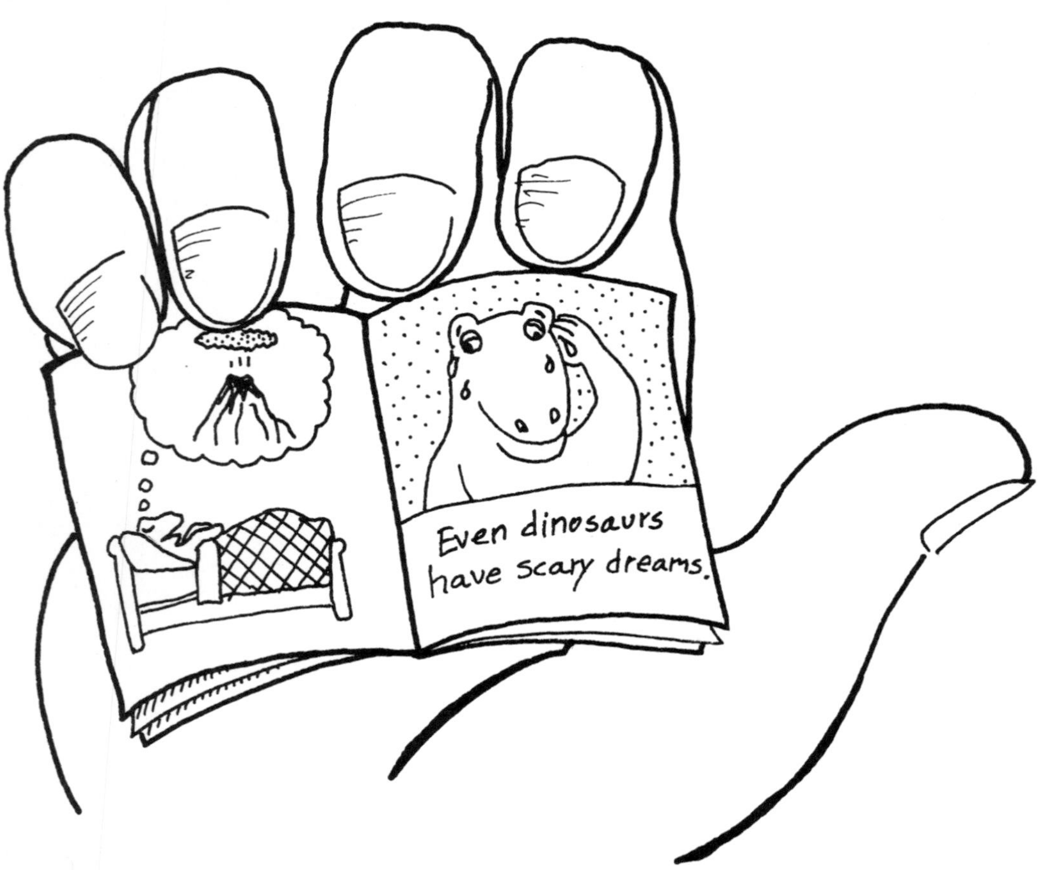
Even dinosaurs have scary dreams.

# TONGUE TWISTERS

**MATERIALS:** paper, writing tools

**DIRECTIONS:**
1. Write a phrase or a sentence that is hard to say. Most of the words should start with the same sound: for example, "Bella bats balloons badly."
2. Test the tongue twister twice.
3. If the tongue twister works, copy it onto a clean piece of paper.
4. Add a silly picture.
5. Collect a bunch of tongue twisters, and make a book of them.

Say several silly sentences speedily!
Pink petuni
public park
Almost every evening Aunt Ellie eats eels.

# TOUCHABLE BOOK

**MATERIALS:** paper, writing tools, glue, touchable things

**DIRECTIONS:**

1. Write a story with things in it for the reader to touch. Examples of things to touch:

   curvy things like paper clips
   fuzzy things like hair made out of cotton
   long things like shoelaces
   metal things like pennies
   paper things like cereal box tops
   round things like bottle caps
   rough things like nail files
   smooth things like silk
   tiny things like grass seeds
   woody things like toothpicks

2. Add drawings. Then glue on the touchable things: for example, a character's hair.
3. When reading the book to a young child, encourage him or her to touch the things on the pages.

Rip Van Winkle looked down and couldn't believe what he saw-
-a beard that reached all the way to his toes!
(Beard made of cotton balls.)

# UN-TV BOOK

**MATERIALS:** paper, writing tools

**DIRECTIONS:**
1. Think up something to do that's more interesting than watching TV.
2. Write a page about it. Tell why the activity is better than watching TV.
3. Add a picture that helps explain what's so good about the activity.
4. Collect a bunch of un-TV activities, and make them into a book.
5. Give the book to someone who watches too much TV. Or place it near the person's TV.

*Variation*: Make a book that shows what people did for entertainment in the days before TV was invented.

# WHO'S WHO RIDDLES

**MATERIALS:** paper, writing tools, dictionary, thesaurus

**DIRECTIONS:**
1. Pick a famous person, place, thing, or event.
2. On scratch paper, make a list of clue words that match the subject's initials. For example, if the subject is Santa Claus, some clue words would be:

   Sled Captain
   Sees Chimneys
   Shouts "Christmas"
   Special Cargo

3. Put the clues in order, with the hardest clues at the top.
4. Write a sentence that lets the reader know whether the subject is a person, place, thing, or event.
5. If the riddle is really hard, add a picture or border that can serve as another clue.
6. Write the answer on the back of the page.
7. Collect a bunch of these riddles and make a book.

*Hint:* Reading a dictionary or a thesaurus can be useful in thinking up clue words.

# WORDLESS BOOK

**MATERIALS:** paper, writing tools, drawing tools

**DIRECTIONS:**
1. Think up a plot or borrow one from an old story.
2. Break the story into parts. Tell each part with a picture but no words.
3. "Read" the book to younger children. Show them each picture, and ask them to tell what they think is going on.

# YEARBOOK

**MATERIALS:** photos of everybody in the class or group, paper, writing tools, tape or glue

**DIRECTIONS:**

1. Make "Who I Am" pages that feature individual photos, with paragraphs written by each person.
2. Make "Group" pages that show and tell about what people did together: for example, putting on a play.
3. Make "Special Places" pages that focus on important areas: for example, the library or the playground.
4. Make "We'll Remember" pages that use pictures, drawings, and words to describe memorable happenings: for example, a big storm or the arrival of a pet.
5. Make "We Learned" pages that talk about spelling words, math skills, facts, books, and lessons.
6. Make "Thank You" pages that give thanks to people who helped make the year great: for example, the janitor and the room parents.
7. Make "Wait Until Next Year" pages that tell about people's hopes and plans for the future.

# YELLOW PAGES

**MATERIALS:** paper, writing tools

**DIRECTIONS:**
1. Make a list of skills that other kids might want to learn: for example, juggling, doing magic tricks, baking cookies, flying a kite, or solving math problems.
2. Write an ad that offers lessons in the skill.
3. Collect ads from other students, and make them into a book.
4. Put a copy of the book in the library.

# Making a Book in Four Steps

Young authors can use the following tips to produce attractive, entertaining, and inexpensive books. This activity will sharpen both writing and art skills. Equally important, it will help develop sound work habits.

Since making books isn't a one-time activity, you might set up a publishing area. It could contain various kinds of paper, scissors, pencils, pens, crayons, glue, yarn, stick-on stars, a stapler, stencils, and a typewriter or a word processor. Keep your eyes open for free—but invaluable—materials, such as unwanted magazines, newspapers, and wallpaper books, which can supply all sorts of pictures, artwork, and lettering. Then dream up a name for the publishing company.

**STEP 1. WRITING THE PARTS OF THE BOOK**

Most books consist of the following parts in this sequence:

front cover
title page
dedication
table of contents
body of the book
author's page
back cover

We'll discuss the parts in this order, though in practice the body of the book will almost always be written first.

**Front Cover**

A catchy title can entice readers to open a book and read. Most successful titles fit into one of the following groups:

Character names: for example, *Barky and the Twins*
Main actions or topics: for example, *Soap Sculpting*
Questions: for example, *Why Not Learn to Whistle?*
Alliteration: for example, *Pigeons, Petunias, and Pigs*
Rhymes: for example: *Rules for Schools*
Commands: for example, *Make and Fly Paper Airplanes*
Numbers: for example, *34 Ways to Avoid Being Bored*
Recycled titles that are variations on famous titles: for example, *Everything You Always Wanted to Know About Hopscotch but Were Afraid to Ask*

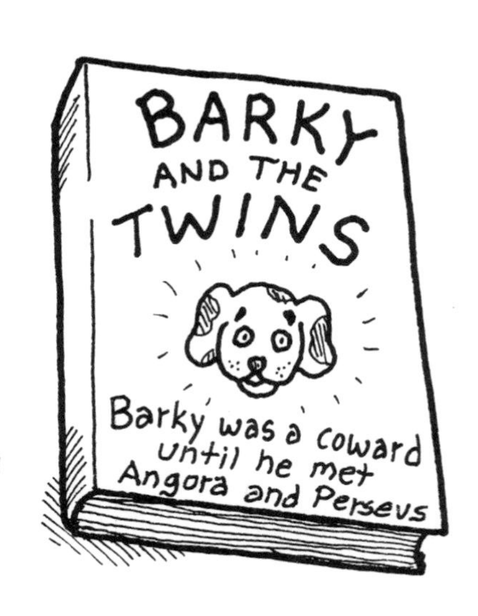

A good way to come up with an outstanding title is to write five or ten possibilities and then choose the best of the bunch. To help make the choice, many authors ask friends for their opinions.

The cover will also name the author or authors, and it often includes a few important words about the contents.

**Title Page**

The title page presents the title, the author's name (or authors' names), and the name of the publisher. If there are many authors, there might be a group name.

**Dedication Page**

The dedication page honors one or a few people admired by the author. Sometimes the dedication note simply mentions a person's name—"For Janet"—but it may also give the reason for the dedication, as with the following example from a class book about special occasions.

> To our parents for giving us the opportunity to have so many Great Moments in our lives.

It's a good idea to give a copy of the book to the person or persons named in the dedication.

While a copyright notice often goes on a separate page, to save paper some publishers put it at the bottom of the dedication page. A typical copyright notice reads like this:

> Copyright 1988 by Tanya James and Melody Myers

**Table of Contents**

A contents page is an important part of any book that has more than one chapter. For example, a book about trees might include chapters on roots, trunks, branches, leaves, fruit, and animals that inhabit trees. In this case, the contents page can help readers find the part of the subject they're looking for.

A contents page can also be useful in a group book. For example, if 30 authors each contribute a page of haiku poems, a contents page listing authors alphabetically can help a reader quickly locate the poetry of a particular poet.

In an anthology, try listing the contents by the type of writing. For example:

Humor . . . . . . . . . . . . . . . . . . . . . . . . . . . . . . . . 5
Riddles . . . . . . . . . . . . . . . . . . . . . . . . . . . . . . . 11
Poems . . . . . . . . . . . . . . . . . . . . . . . . . . . . . . . . 17

**Body**

The body is the main part of the book. In an ABC book, it might consist of one page for each letter of the alphabet. In a storybook, the body includes all the pages that tell the story.

Authors start by writing a rough draft, often on scratch paper. Improving the draft may require crossing out, adding, and moving words. It may help to share the work with someone and ask for suggestions to make the wording clearer and more interesting.

When everything sounds right, check each page for spelling and punctuation.

**Author Page**

Readers are interested in learning about the author or authors. Include facts that answer questions like:

How old is the author?
What are the author's hobbies?
Why did the author write this kind of book?
What other books has the author written?

If the book is a collection of pages by many writers, then the authors' page might focus on the group. For example, the authors' page for a riddle collection by 30 students might read:

This book represents the writing of all the fourth graders in Room 151. This group of authors—known as The Writing Regulars—has produced six other group books on such topics as sports and dinosaurs.

**Back Cover**

The outside back cover may list more details about the book. It might even contain a passage from the book.

## STEP 2. CREATING THE PAGES

### Design and Layout

Plan the look of each page and the covers. This starts with choosing the kind of art—decorative or content-oriented. Next, decide how big the art should be and where it will go. For example, will the words always be on the left side of the page and the art on the right? Or will each page look different?

This step often involves making a sketch. Wavy lines represent words; boxes indicate the art. Sketching is a big help in a group project if the pages are meant to resemble each other. Give extra care to the cover art. Often one big, colorful picture will have more impact than several small pictures.

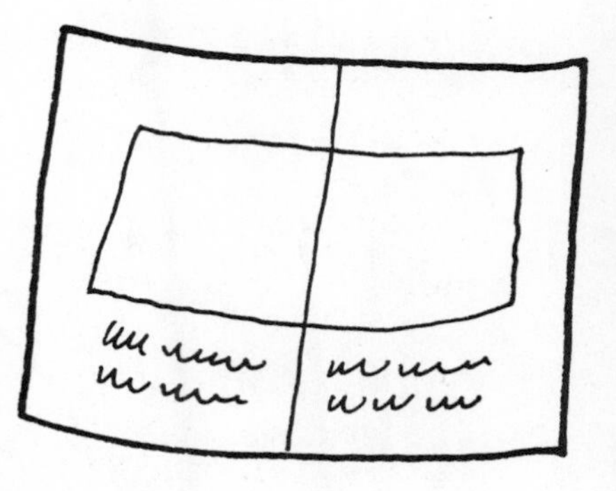

### Paper

For books that consist of art and words, plain white typing or ditto paper works best. Occasionally, you might experiment with using colored paper: for example, a difer-ent color for each type of writing in an anthology.

For the sake of durability, make the cover out of double-layered construction paper, cardboard, or student artwork backed with construction paper. Other possibilities are poster board, oaktag, or felt. Laminating or using clear adhesive shelf paper on the covers makes them resistant to stains and wear.

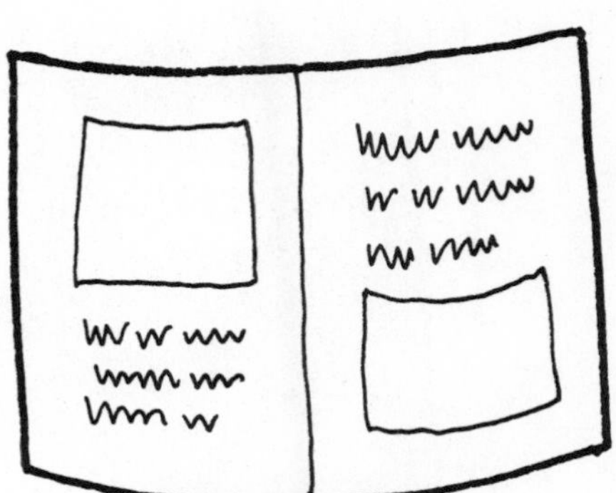

### Lettering

Typewritten or computer-printed books look real. But hand-lettered volumes can also be attractive. To obtain straight lines, place lined paper under a blank sheet. For pizzazz, print titles, chapter headings, or key words in color. Another technique is to begin the first word of a page or a chapter with a large "illuminated" letter.

## ILLUMINATED LETTERS

A big, first letter on a page can dazzle readers. Such letters can be drawn:

or cut from a picture:

or made of stick-on stars:

**Illustrations**

Illustrations can include original drawings or photographs, glued-on objects (for example, an object ABC book), or artwork clipped from magazines and newspapers. Publishers may also photocopy pictures from books of ready-made illustrations. You can find examples in art supply stores, or write: Dover Publications, 31 East 2nd Street, Mineoloa, New York 11501. Similar ready-to-use illustrations are available on computer disks. Another approach is to have a friend or relative provide the drawings, just as many children's book authors team with artists.

Borders can give a book unity. This is especially important for group projects.

*Hint:* If multiple copies are desired, keep the artwork relatively simple. Make a trial copy to see how well it reproduces on a copy machine.

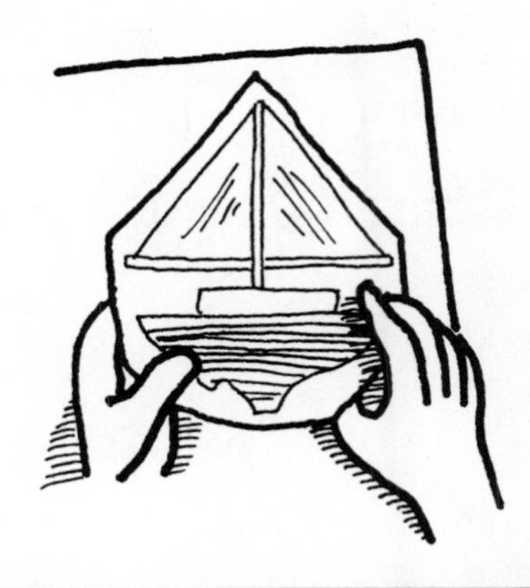

## BORDERS

Borders add interest to a book.

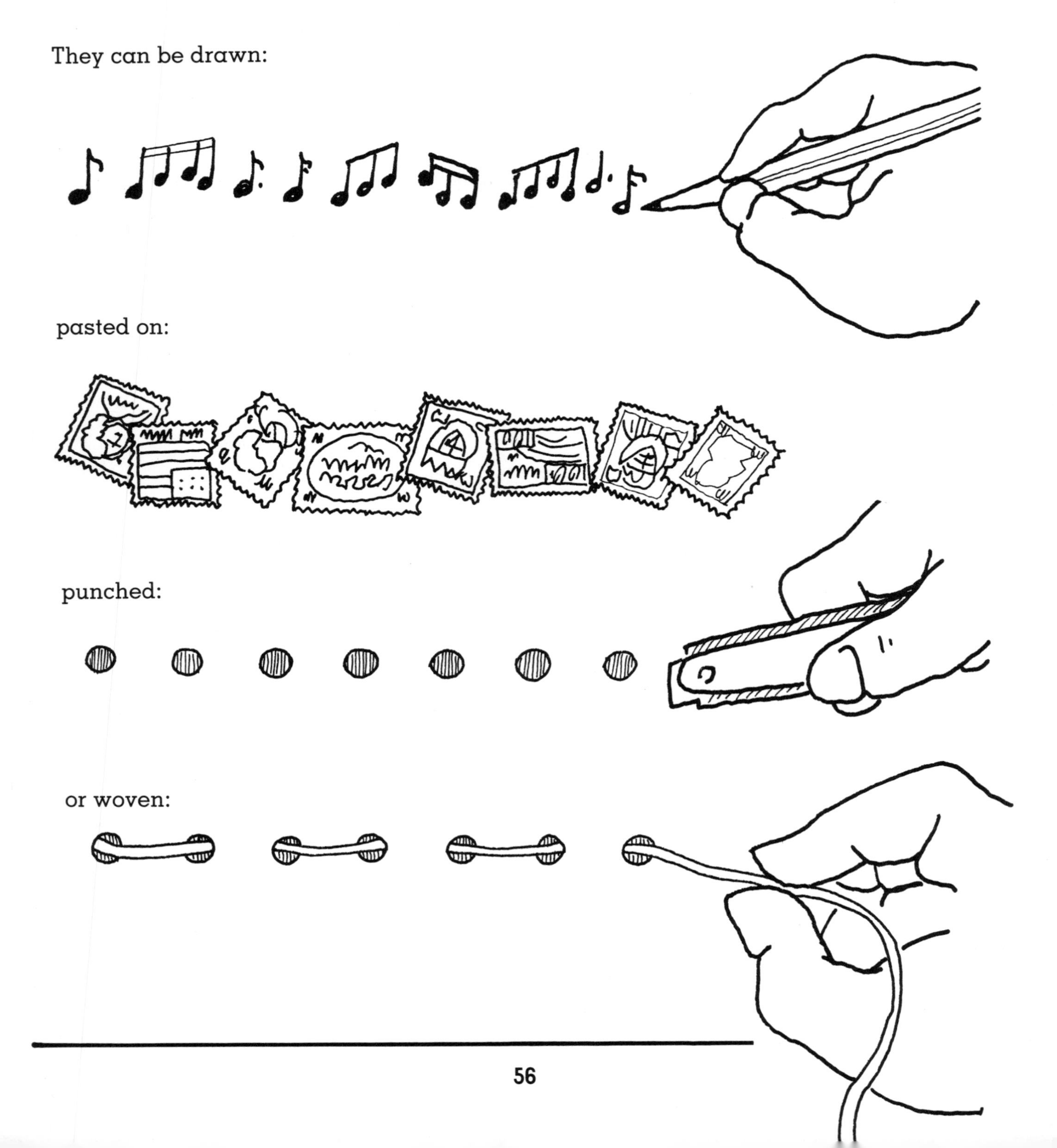

They can be drawn:

pasted on:

punched:

or woven:

While borders can be just for decoration:

they can also relate directly to the subject. For example, a page about money can have a dollar-sign border:

$ $ $ $ $ $ $ $ $ $ $ $ $ $ $ $

while the border of a riddle book might look like this:

??????????????????????????

## STEP 3. BINDING THE PAGES

Here are five ways to gather pages into a book.

**Stapling**

The simplest and least expensive technique is to staple the pages and covers along the edge. Place colorful cloth tape over the edge to cover the staples and to reinforce the binding.

**Binder**

The second method is to use a ready-made plastic or paper binder. Some use slides or metal clasps. With those that require hole-punching the pages, it's a good idea to reinforce the holes.

**Rings, Yarn, Ribbon, or Twine**

Hole-punch each page, reinforce the holes, and bind with metal rings, braided yarn, ribbon, or twine. Score the cover so that the pages will fold back easily.

**Spiral Binding**

This method requires use of a GBC machine available in many school offices. Some photocopy stores will spiral-bind books. They may also offer an alternative called Velo binding for about the same price.

**Stitched Binding**

This approach involves copying or pasting the completed words and art into a book of blank pages. Such blank books can be purchased in book or stationery stores. Another source is Barebooks, Treetop Publications, 220 Virginia Street, Racine, WI 53405.

You can also make your own stitched book. Begin by preparing a model or dummy book. Every page in the finished book—back and front, including the covers—must be represented in the dummy. The pages in the dummy can simply be stapled.

Next, make a stack of paper twice as wide as the pages in the finished book. To calculate the number of sheets you'll need, divide the pages in the dummy by four. For example, if the dummy has 32 pages, you'll need 8 sheets of large paper. (Always round upward. For example, if there were 37 pages, 37/4 = 9 1/4 which should be rounded up to 10.)

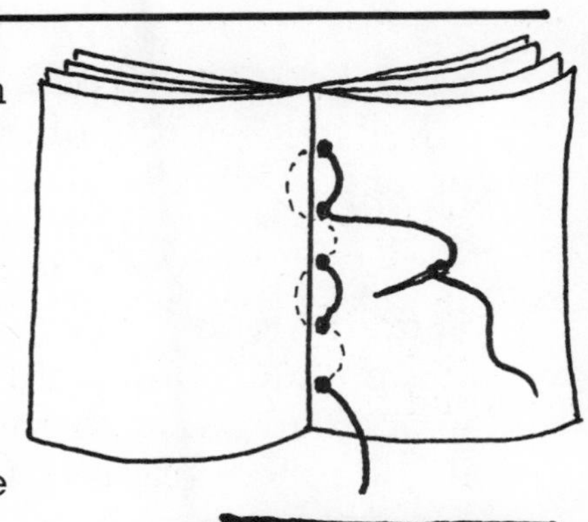

• Carefully fold the pages in half to make a crease. Then unfold the stack. Note that each folded sheet will make four pages.

• Use a heavy-duty needle to poke five holes along the fold.

• Stitch from the outside spine to the center. Go up and over the row of holes at least twice. Knot the thread or tie it.

• Cut two pieces of cardboard slightly bigger than the pages.

• Cut two pieces of fabric slightly larger than the card-board pieces.

• Glue each piece of cardboard to the inside of the fabric. Glue the corners to the cardboard, and then glue the edges.

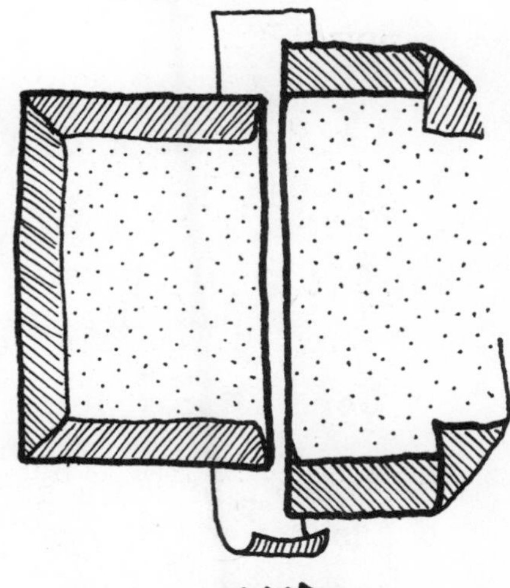

• Cut a strip of tape that's longer than the boards. Lay the tape sticky-side up. Set the boards about the width of a pencil apart onto the tape. Bring the top and the bottom tails of the tape over the boards and fold down. This is the spine of the book.

• Cover the spine with a strip of paper and put the pages into the center of the spine.

• Glue the first page to the left board. Then glue the last page to the inside of the right board. Hold the book pages at a right angle when gluing.

• For end papers, cut two pieces of paper and glue them onto the inside front and back covers.

Now, using the dummy as a guide, copy each page from the draft into the book.

## STEP 4. SHARING THE BOOK

Just as the proof of the pudding is in the eating, the proof of a book is in the reading. Here are four ways to get a book into circulation.

### Read It Aloud

A popular forum is the story hour where an author reads to one child or to a group. This might occur in a classroom, in the school library, in an assembly, or in the public library. Local bookstores also might sponsor an authors' night.

If a video or an audio recorder is handy, readings might be taped for sharing at home or with groups elsewhere, such as at another school.

Books might also be shared at a Book Festival or Young Authors' Conference held, for example, as part of a school open house or a celebration of National Library Week. In addition to young writers, presenters might include local authors.

### Display It

Share books through displays in hallways, libraries, or the windows of local shops. Include a poster that gives some background on the publishing activity.

If the authors are willing, originals or copies might be sold in order to raise funds for charity or for buying materials to publish more books.

### Donate It to a Library

A classic way to share a book is to place it in a library. This might be the school library, the town library, or a classroom library.

If the book goes into the classroom library, compose catalog cards to create a class card catalog. Use index cards and a file box for storage.

While modern libraries are switching to computerized catalogs, catalog cards work fine for classroom collections.

To prepare children for the future, the examples below use a popular computer format rather than the traditional layout. However, there's one difference between these cards and what the computer shows. To enable kids to sort the cards, we've added a first line that gives either the author, the title, or the subject.

**Author Card**

Johnson, Laura

| | |
|---|---|
| CALL # | 793 |
| AUTHOR | Johnson, Laura |
| TITLE | Can You Guess It? |
| PUBLISHED | St. Louis, MO: School House Books. 1987 |
| DESCRIPTION | 16 pages illustrated |
| SUMMARY | Presents humorous riddles. |
| TOPICS | 1) Animal riddles 2) Sports riddles |

**Title Card**

Can You Guess It?

| | |
|---|---|
| CALL # | 793 |
| AUTHOR | Johnson, Laura |
| TITLE | Can You Guess It? |
| PUBLISHED | St. Louis, MO: School House Books. 1987 |
| DESCRIPTION | 16 pages illustrated |
| SUMMARY | Presents humorous riddles. |
| TOPICS | 1) Animal riddles 2) Sports riddles |

**Subject Card**

Riddles

| | |
|---|---|
| CALL # | 793 |
| AUTHOR | Johnson, Laura |
| TITLE | Can You Guess It? |
| PUBLISHED | St. Louis, MO: School House Books. 1987 |
| DESCRIPTION | 16 pages illustrated |
| SUMMARY | Presents humorous riddles. |
| TOPICS | 1) Animal riddles 2) Sports riddles |

## Give It Away

Few gifts are more appreciated than a book, and few gift books more treasured than a hand-made book. As a bonus, include a bookmark with the book.

Bookmarks—made out of construction paper or cloth—are a great gift for any reader. The following examples are meant to spark other ideas.

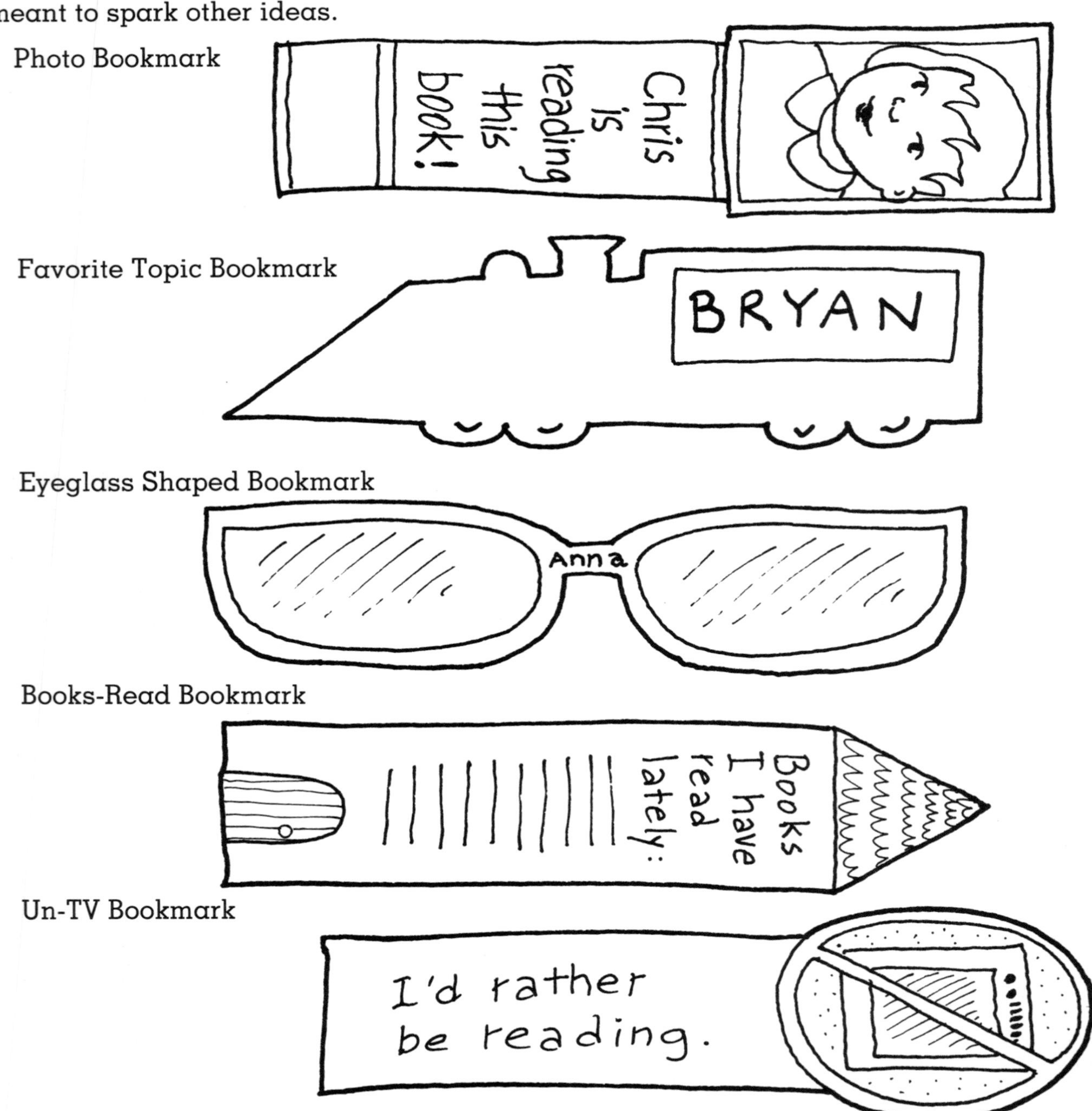

## ART IDEAS

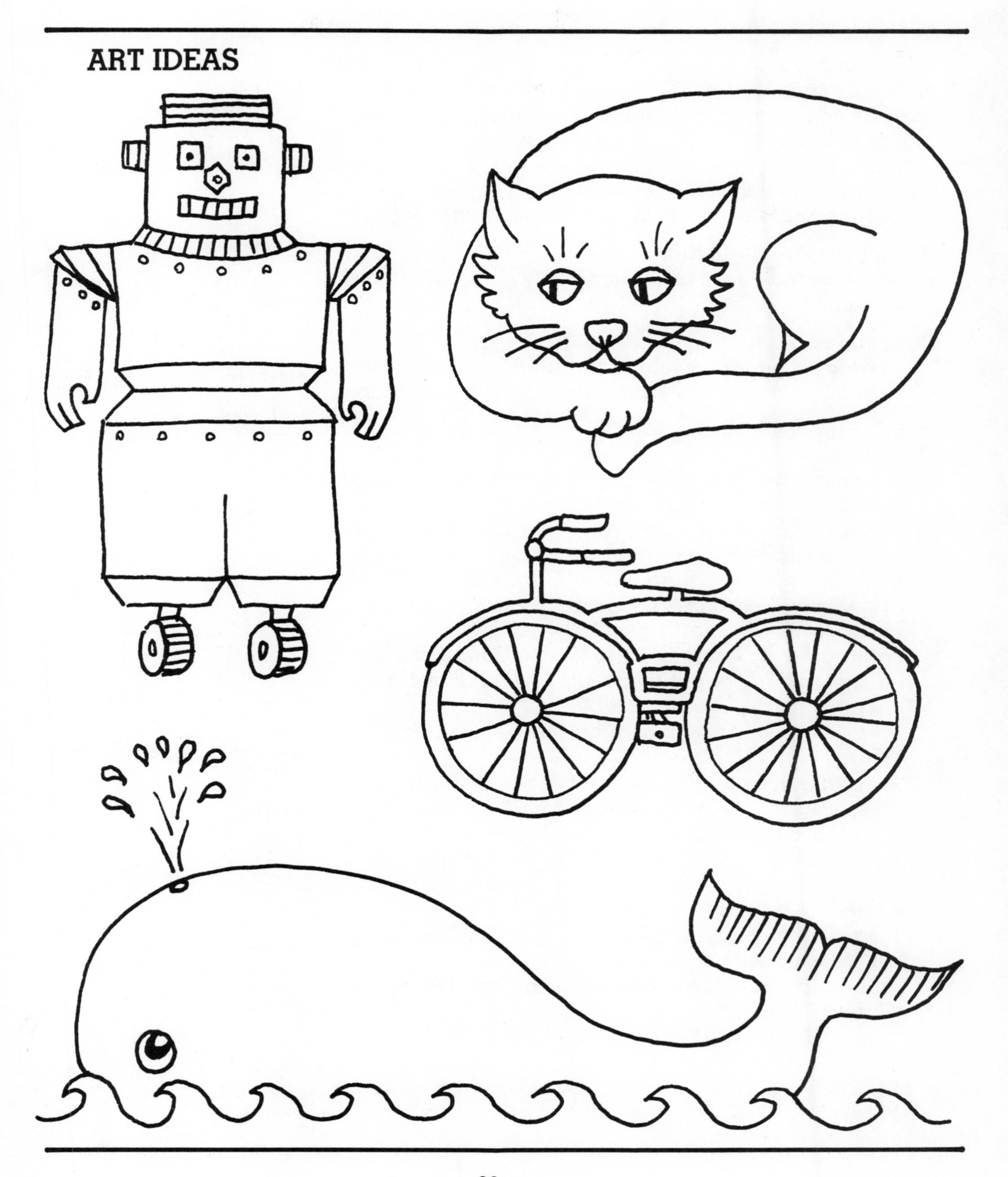

# Book List

One of the best ways to teach any writing assignment is to start with models. The following books present professional examples of some of the more unusual formats described in *The Book Factory*.

**Choose-the-Ending Stories**

*Choose Your Own Adventure Series* (Bantam Skylark)

**Math Books for Little Kids**

*Bunches and Bunches of Bunnies* by Louise Matthews (Scholastic, 1978)

**Moveable Books**

*Lavinia's Cottage* by John Goodall (Atheneum, 1983)
*Barbara's Birthday* by James Stevenson (Greenwillow, 1983)

**Pattern Books**

*A House Is a House for Me* by Mary Ann Hoberman (Viking, 1978)
*Animals Should Definitely Not Wear Clothing* by Judi Barrett (Atheneum, 1970)
*Dark, Dark Tale* by Ruth Brown (Dial, 1981)
*Fifty-Seven Reasons Not to Have a Nuclear War* by Marty Asher (Warner Books, 1984)
*Fortunately* by Remy Charlip (Macmillan, 1985)
*I Know an Old Lady Who Swallowed a Fly* by Nadine Westcott (Little, Brown, 1980)
*If I Found a Wistful Unicorn* by Ann Ashford (Peachtree, 1978)
*Important Book* by Margaret Wise Brown (Harper and Row, 1949)
*Napping House* by Audrey Wood (Harcourt, 1984)
*Q Is For Duck* by Mary Elting (Houghton Mifflin, 1980)
*Some Things Are Scary* by Florence Heide (Scholastic, 1969)
*What Do You Say, Dear?* by Sesyle Joslin (Young Scholastic Books, 1958)
*Where Does the Sun Go At Night?* by Mirra Ginsburg (Greenwillow, 1981)
*Whose Mouse Are You?* by Robert Kraus (Macmillan, 1970)
*Would You Rather...* by John Burningham (Crowell, 1978)